ROOTS

by Melanie Mitchell

first step nonfiction

Lerner Publications Company · Minneapolis

Look at the roots.

Most plants have roots.

Roots are in the ground.

Plants use roots to get food.

Roots can be big.

Roots can be small.

Do you see roots?